Math Concepts

Doubles Fun
on the Farm

Copyright © Gareth Stevens, Inc. All rights reserved.

Developed for Harcourt, Inc., by Gareth Stevens, Inc.
This edition published by Harcourt, Inc., by agreement with Gareth Stevens, Inc. No part of this publication may be reproduced or transmitted in any form or by any means, electronic or mechanical, including photocopy, recording, or any information storage and retrieval system, without permission in writing from the copyright holder.

Requests for permission to make copies of any part of the work should be addressed to Permissions Department, Gareth Stevens, Inc., 330 West Olive Street, Suite 100, Milwaukee, Wisconsin 53212. Fax: 414-332-3567.

HARCOURT and the Harcourt Logo are trademarks of Harcourt, Inc., registered in the United States of America and/or other jurisdictions.

Printed in Mexico

ISBN 13: 978-0-15-360228-3
ISBN 10: 0-15-360228-7

4 5 6 7 8 9 10 0908 16 15 14 13 12
4500361024

Doubles Fun
on the Farm

by Joan Freese
Photographs by Russell Pickering

Chapter 1: To the Farm

Max lives in the city. He plans to visit his cousin. Her name is Sara. Sara lives on a farm.

Max loves to go to the farm. He visits often. He likes to play with Sara. He has so much fun.

Chapter 2:
Finding Doubles at the Farm

Max's dad takes him to Sara's house. Max is excited. "Hello, Sara!" he shouts.

"Hi, Max!" Sara says. Just then Max sees a bird.

$1 + 1 = 2$

Sara spies another bird. "1 bird plus 1 bird equals 2 birds," she says. "That is a double! A double is when you add two numbers that are the same."

Sara has two dogs. They like to run. They like to play. Sara and Max play fetch with them. Max thinks about doubles.

$$2 + 2 = 4$$

"I get it," he says. "Each dog has 2 ears. 2 dog ears plus 2 dog ears equals 4 dog ears. That is a double, too."

Sara and Max like playing outdoors. Being outside makes them hungry. Soon it is snack time. Sara's mom brings them something to eat.

$3 + 3 = 6$

The cousins have crackers for a snack. Sara takes 3 crackers. Max takes 3 crackers. They have 6 crackers in all. A double again!

Now the cousins go for a walk. Sara's mom goes, too. They hunt for four-leaf clovers. The clover plants are in the grass.

$4 + 4 = 8$

Sara looks and looks. She finds two plants. Each has 4 leaves! 4 leaves and 4 leaves make 8 leaves. Another double!

Max and Sara head to the stream. The stream is on Sara's farm. They take off their shoes. They wade in the water.

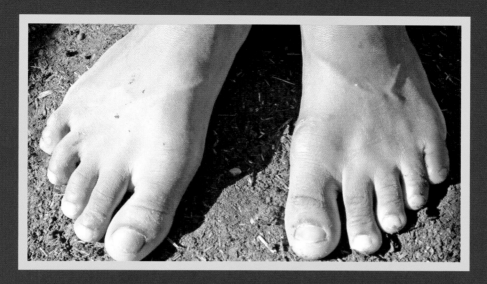

$5 + 5 = 10$

"Look, my toes are a double!" Max says.
"I have 5 toes on each foot."

Sara looks at her toes. "Mine too," she says.

Chapter 3:
More Doubles Fun

Later, Sara shows Max the garden. Beans and corn grow there. Other food grows in the garden, too. Max likes to eat corn.

$6 + 6 = 12$

Sara likes carrots best. She pulls 6 carrots from the ground. Max pulls out 6 carrots, too. They take 12 carrots to Sara's mom.

Next the cousins visit the chickens. Max and Sara each find a basket. They fill their baskets with eggs.

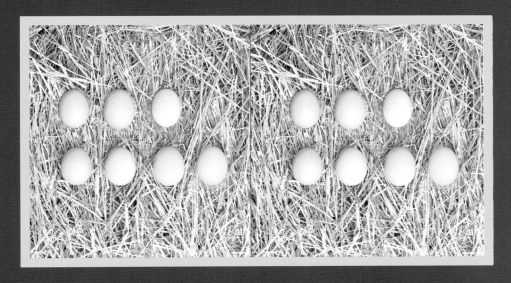

Sara has 7 eggs in her basket. Max counts out 7 eggs, too. More doubles fun!

Sara and Max walk to the house. Sara sees two spiders. The spiders are making webs. "Look, a spider city!" says Sara.

$8 + 8 = 16$

Sara and Max look again. Each spider has 8 legs. Max thinks about doubles. "8 legs and 8 more legs makes 16 legs in all!"

Chapter 4:
One Last Double

It is almost time for Max to go home. The cousins rest near some trees. The branches hold many apples.

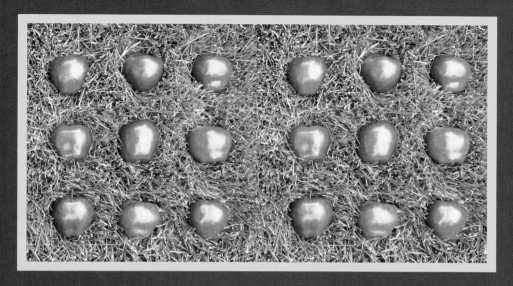

$9 + 9 = 18$

Max picks 9 apples. Sara picks 9 apples, too. Max will bring all 18 apples home. They will remind him of the doubles fun today.

Glossary

add to join two groups

doubles an addition fact in which both addends are the same. 8 + 8 = 16 is a doubles fact.

equals has the same amount or value

farm land used to grow crops or raise animals

plus added to

stream a small body of flowing water

Photo credits: cover, title page, pp. 3, 6, 8, 9, 10, 12, 13, 15, 16, 17, 18, 19, 20, 21 Russell Pickering; p. 5 National Biological Information Infrastructure; p. 7 © Pinto/zefa/Corbis; p. 11 © Josh Westrich/zefa/Corbis.